Mouaad Amine Mazri

Auxins: Roles and functions in plant tissue culture

Mouaad Amine Mazri

Auxins: Roles and functions in plant tissue culture

Noor Publishing

Imprint

Any brand names and product names mentioned in this book are subject to trademark, brand or patent protection and are trademarks or registered trademarks of their respective holders. The use of brand names, product names, common names, trade names, product descriptions etc. even without a particular marking in this work is in no way to be construed to mean that such names may be regarded as unrestricted in respect of trademark and brand protection legislation and could thus be used by anyone.

Cover image: www.ingimage.com

Publisher:
Noor Publishing
is a trademark of
International Book Market Service Ltd., member of OmniScriptum Publishing Group
17 Meldrum Street, Beau Bassin 71504, Mauritius

Printed at: see last page
ISBN: 978-3-330-97365-7

Copyright © Mouaad Amine Mazri
Copyright © 2020 International Book Market Service Ltd., member of OmniScriptum Publishing Group

Auxins

Roles and functions in plant tissue culture

Mazri Mouaad Amine

Institut National de la Recherche Agronomique,

CRRA-Marrakech, UR Agro-Biotechnologie,

BP 533, Marrakech, Morocco.

e-mail: m.a.mazri@gmail.com

Auxins are a group of plant hormones and growth regulators that are synthetized naturally by plants or manufactured artificially, and that play a major role in plant micropropagation.

Auxins significantly influence cell division, growth, differentiation and in vitro morphogenesis. This is due to the fact that exogenous auxins interact with endogenous plant hormones and modify their concentrations. As a result, morpho-physiological responses of plant cells and tissues cultured in vitro are observed.

Auxins are also known for their role in maintaining apical dominance and plant polarity, and are involved in the plant signaling systems, which allows plant cells to interact among them in response to different exogenous factors and culture conditions.

Based on their crucial role in plant morphogenesis, and their significant impact on the different developmental processes of plant cells and tissues cultured in vitro, auxins have been widely used to induce rhizogenesis, caulogenesis, callogenesis and somatic embryogenesis, and thus have a major importance in plant micropropagation.

The present book gives a definition and historical overview of auxins. It presents the different natural and synthetic auxins and describes their roles and modes of action, and reports results from literature that highlight the significant impact of auxins in plant micropropagation.

2,4-D	2,4-dichlorophenoxyacetic acid
2,4,5-T	2,4,5-trichlorophenoxyacetic acid
2iP	6-(γ,γ-Dimethylallylamino)purine
BAP	6-benzylaminopurine
BSAA	3-benzo[b] selenienyl acetic acid
Dicamba	3,6-dichloro-2-methoxybenzoic acid
DMF	Dimethylformamide
DMSO	Dimethylsulfoxide
ESM	Embryogenic suspensor mass
GA$_3$	Gibberellic acid
GA$_4$	Gibberellic acid 4
IAA	Indole-3-acetic acid
IBA	Indole-3-butyric acid
NAA	1-naphthaleneacetic acid
NOA	2-naphthoxyacetic acid
p-CPA	*p*-chloro-phenoxyacetic acid
PEG	Polyethylene glycol
PGR	Plant growth regulator
Picloram	4-amino-3,5,6-trichloro-2-pyridinecarboxylic acid
PA	Polyamine
TDZ	Thidiazuron

TABLE OF CONTENT

INTRODUCTION

Auxins are a group/family of plant hormones and growth regulators that are synthetized naturally by plants or produced artificially, and that play a major role in in vitro propagation of plants.

Auxin is a term used to describe all the natural and synthetic compounds that have a similar action to indole-3-acetic acid such as inducing cell division, growth and elongation. It is a Greek origin word that means 'to grow' or 'to enlarge'. Since their discovery, auxins have been widely used in plant cell, tissue and organ culture. In fact, auxins are involved in cell division and growth, which results in morphogenesis.

In addition to cell division, growth and elongation, auxins are mainly known for their potent impact in inducing rhizogenesis.

In plant tissue culture and in vitro morphogenesis, auxins are generally used in combination with another group of plant growth regulators: cytokinins. In fact, these two groups have a complementary role in cell division and growth, and in regulating in vitro morphogenesis. However, in many plant species, auxins may be used alone. For example, to induce rhizogenesis or callogenesis. Auxins are also known for their role in maintaining apical dominance and plant polarity.

Auxins may be naturally synthetized by plants, or prepared artificially. The auxins that are naturally synthetized by plants are known as endogenous auxins and are used at very low amounts. They have a growth regulatory role in plant development rather than a nutritional role. On the other, some chemical compounds with the same roles and functions as endogenous auxins are synthetized artificially. These compounds are also used at very low concentrations to regulate plant growth and development and are known as exogenous auxins. The exogenous auxins are added exogenously to the culture medium and have the same biological activity as the endogenous ones. However, it is generally believed that exogenous auxins are more potent than those synthetized naturally by plants.

Auxins and cytokinins, when added in combination to the culture medium, significantly influence cell division, growth and in vitro morphogenesis. This is due to the fact that exogenous auxins and cytokinins interact with endogenous plant hormones and modify their concentrations. As a result, a morpho-physiological response of plant cells and tissues cultured in vitro is observed.

Auxins are also involved in the plant signaling systems, which can be defined as communication and coordination systems that allow plant cells to interact among them as a response to the different external factors and conditions. This interaction guides the different plant cycles of growth and development. Auxins, when added to plant cells and tissues cultured in vitro, get involved in plant signaling systems and thus play a major role in many physiological and biochemical processes of plant growth, development and morphogenesis.

Adding auxins to culture medium makes DNA more methylated than in a normal state. This methylation results in the re-programming of differentiated cells. High DNA methylation results in the induction of many morphogenetic responses. For example, somatic embryogenesis.

Based on their major role in plant morphogenesis, and their significant impact on the different developmental processes of plant cells and tissues cultured in vitro, auxins have been widely used to induce rhizogenesis, caulogenesis, callogenesis and somatic embryogenesis, and thus have a high importance in plant regeneration in vitro.

The concentration and type of auxin to use for plant micropropagation depend strongly on the genotype, the explant type and the micropropagation technique used.

The present book starts by giving a definition and historical overview of auxins. It presents the different natural and synthetic auxins and describes their roles and modes of actions. Finally, it reports examples from literature that highlight the significant impact of auxins in in vitro regeneration of some plant species by employing techniques such as somatic embryogenesis, organogenesis and microcuttings.

DEFINITION AND HISTORICAL OVERVIEW OF AUXINS

DEFINITION OF AUXINS

Before defining the term '*auxin*', it is important to introduce and define first some other terms that are generally linked to auxins. The first one of them is '*indoleacetic acid*':

❖ Indoleacetic acid could be defined as a plant hormone synthetized by plants and involved in various biochemical processes of plants including cell division, growth and differentiation (Van Overbeek 1959).

Other terms such as plant regulators, plant hormones, growth regulators, growth hormones, flowering regulators and flowering hormones should also be defined. According to 'The American Society of Plant Physiologists' (Van Overbeek 1959), these terms could be defined as follows:

❖ Plant regulators are organic compounds, generally used in small concentrations in order to stimulate, prevent or change physiological and biochemical processes involved in plants growth and development.

❖ Plant hormones, also known as phytohormones, are a class of compounds naturally synthetized by plants. They are involved in the physiological and biochemical processes of plants and thus are important for plant growth and development. Generally, phytohormones act in a site different to that in which they are synthetized.

❖ Growth regulators, also known as growth substances, are a group of compounds that influence plant growth.

- ❖ Growth hormones refer to all the compounds that are involved in plant growth and development.
- ❖ Flowering regulators are the compounds involved in regulating plant flowering.
- ❖ Flowering hormones refer to the hormones involved in floral primordia initiation, growth and development.
- ❖ Auxin precursors refer to the compounds that can be converted into auxins.
- ❖ Anti-Auxins refer all the compounds that have an inhibitory effect on auxin actions.

Auxin is a term used to describe all the natural and synthetic compounds that have a similar action to indole-3-acetic acid (IAA) such as inducing cell division, growth and elongation (Van Overbeek 1959). It has a Greek origin that means to grow or to enlarge (Machakova et al. 2008). Auxins are involved in cell division and growth, which results in the formation and proliferation of unorganized cells (callogenesis) and defined organ (morphogenesis; Machakova et al. 2008).

In addition to cell division, growth and elongation, auxins are known to induce rhizogenesis.

Generally, all auxins are commercially known as root-promoting compounds. However, the naturally occurring auxins such as IAA and other endogenous hormones are scientifically known as plant hormones. On the other hand, synthetic auxins are generally known as plant growth regulators (Blythe et al. 2007).

Auxins are generally acids with an unsaturated cyclic nucleus. They have a similar physiological reaction pattern as indoleacetic acid (Van Overbeek 1959). This is the main reason why they are distinct from other plant growth regulator (PGR) groups such as cytokinins and gibberellins, which are also involved in cell division, growth and elongation of cells.

In plant tissue culture and in vitro morphogenesis, auxins and cytokinins are the most frequently used groups of PGRs. Auxins and cytokinins are generally used in combination in order to induce and stimulate callogenesis and to regulate in vitro morphogenesis (Machakova et al. 2008). Auxins are also known for their role in maintaining apical dominance and plant polarity (Friml 2003).

DISCOVERY AND HISTORICAL OVERVIEW OF AUXINS

According to Van Overbeek (1959), the article of Darwin entitled '*The Power of Movement in Plants*' published in 1880 could be considered as the origin of investigations on plant hormones and plant growth regulators. In his article, Darwin reported the existence of a stimulus inside plants that is transmitted from one part to an adjoining part.

In 1928, Went (1928) reported that Darwin was describing plant hormones.

In 1934, Kogl et al. (1934), succeeded to chemically separate and identify indoleacetic acid as an auxin. Few years later, Chadwick and Kiplinger (1938) reported that auxins improve rooting rates and quality. However, such impact depends strongly on plant species. Subsequently, many manufacturers started producing artificial products that contain auxins to be used to accelerate plant rooting.

In 1935, Thimann and Koepfli (1935) described the artificial establishment of IAA and proved its role in stimulating plant rhizogenesis. In the same year, indole-3-butyric acid (IBA) was reported as a homolog of IAA. Soon after, 1-naphthaleneacetic acid (NAA) was reported to have a strong auxin effect (Zimmerman and Wilcoxon 1935). Zimmerman and Wilcoxon (1935) demonstrated that these two synthetic PGRs, NAA and IBA, are more effective in promoting rhizogenesis than IAA.

In 1942, Zimmerman and Hitchcock (1942) introduced 2,4-dichlorophenoxyacetic acid (2,4-D), an auxin highly different from IAA and with a stronger action.

In 1993, Ludwig-Müller et al. (1993) indicated that IBA is naturally synthetized by plants from IAA, but in very low amounts.

NATURAL AND SYNTHETIC AUXINS

Auxins may be naturally synthetized by plants, or prepared artificially. The auxins that are naturally (endogenously) synthetized by plants are generally called plant hormones. They are used at very low amounts and have a growth regulatory role in plant development rather than a nutritional role (Machakova et al. 2008). On the other hand, some chemical compounds with the same roles and functions as endogenous auxins are synthetized artificially. Such compounds are also used at very low concentrations to regulate plant growth and development and in the specific case of auxins, they are known as exogenous auxins (added exogenously to the culture medium). Exogenous auxins have the same biological activity as the endogenous ones, but generally the exogenous auxins are more potent than the endogenous ones (Machakova et al. 2008).

NATURAL AUXINS

Indole-3-acetic acid is by far the most naturally occurring auxin in plants. In 1935, Thimann and Koepfli (1935) described the artificial establishment of IAA and proved its role in stimulating plant rhizogenesis. IAA has been widely used in plant tissue culture and showed potent effect on callogenesis, organogenesis and somatic embryogenesis. It can be used alone in the culture medium but is generally used in combination with a cytokinin to induce morphogenesis. The concentration used of IAA is also known to have a significant impact on explant response. Generally speaking, high concentrations of IAA induce callogenesis while low concentrations of this auxin induce direct organogenesis (caulogenesis and rhizogenesis). However, such responses depend on plant species, genotype and the explant in culture.

Indole-3-butyric acid and 4-chloro-IAA were also reported to be naturally synthetized by some plants, as a product resulting from IAA. However, IBA and

4-chloro-IAA are synthetized at very low concentrations (Ludwig-Müller and Epstein 1994; Ludwig-Müller et al. 1993; Engvild 1985).

SYNTHETIC AUXINS

In a wide range of plant species and genotypes, 2,4-D, IBA and NAA are the most frequently used synthetic auxins for callus induction, somatic embryogenesis and rhizogenesis. Generally, auxins are used in combination with cytokinins. However, due to the potent role of auxins in inducing calli and roots, some synthetic auxins have been widely used alone to reach those purposes. Auxins were also reported to accelerate root induction, to increase the number of adventitious roots and to improve the overall root system quality (Hartmann et al. 2002; Macdonald 1987).

Other synthetic auxins such as 2-naphthoxyacetic acid (NOA), 4-amino-3,5,6-trichloro-2-pyridinecarboxylic acid (picloram) and 3,6-dichloro-2-methoxybenzoic acid (dicamba) have been also used for the in vitro multiplication of many plant species. Their ability to induce callogenesis, somatic embryogenesis and rhizogenesis depends on plant species, genotypes and explant type.

Synthetic auxins are commercially available in different formulations. For example, as a powder, in liquid state, as a gel or as a water-soluble tablet (Hartmann et al. 2002). However, the powder formulation is the most frequently commercialized one, and different pack sizes are available.

The powder formulation of auxins can be used for up to two years when the auxin is prepared at a low concentration while it can be used for up to five years when the concentration of the auxin is high (Lowenfels 1966). Before utilization, the powder may be dissolved in different types of solvent such as water, ethyl alcohol or methanol, depending on the auxin itself. After preparation, the solution should be stored at 4°C in dark glass bottles.

The liquid formulation can be stored for up to six months at room temperature. However, it is highly suggested to store it in refrigeration (at 4° C) in dark glass bottles. Besides, changes in color during the storage of the liquid formulation may occur. Nevertheless, no significant change in the biological activity or the effectiveness of auxins with changed color was reported (Robbins et al. 1988). Ethyl alcohol and isopropyl alcohol are the most commonly used solvents in liquid formulations (Dirr et al. 1987). Other solvents such as polyethylene glycol, propylene glycol, methanol, acetone, dimethylsulfoxide (DMSO), dimethylformamide (DMF) and ethylene glycol have been also used (Amissah and Bassuk 2004; Chong and Hamersma 1995 a,b; Dirr 1981; Howard et al. 1984).

MODE OF ACTION OF AUXINS

Skoog and Miller (1957) were among the first scientists to highlight the importance of combining auxins and cytokinins for plant micropropagation and their significant role in promoting the main processes behind in vitro organogenesis. Nowadays, it is well known that auxins and cytokinins, when added in combination to the culture medium, significantly influence cell division, growth and in vitro morphogenesis. According to Gaspar et al. (1996), this is possible as exogenous auxins and cytokinins interact with endogenous plant hormones and modify their concentrations. As a result, a morpho-physiological response of plant cells and tissues cultured in vitro is observed.

According to Libbenga and Mennes (1995), auxins, with other plant hormones, are involved in the plant signaling systems. Plant signaling systems could be defined as communication and coordination systems that allow plant cells to interact among them as a response to the different external factors and conditions. This interaction guides the different plant cycles of growth and development. Auxins, when added to plant cells and tissues cultured in vitro, get involved in plant signaling systems and thus play a major role in many physiological and biochemical processes of plant growth, development and morphogenesis.

As mentioned above, auxins, when used alone or in combination with other PGRs such as cytokinins, are generally known to impact cell division, multiplication, growth and elongation. According to Stickens et al. (1996), the auxin and cytokinin combination added to the culture medium influence cell division, multiplication, growth and elongation by controlling the required signal for in vitro morphogenesis.

According to Ljung (2014), auxin is a highly important signaling molecule that use specific receptors, the most known class of these receptors is the F-box class, also known as Transport Inhibitor Response/Auxin Signaling F-Boxes (TIR1/AFBs). The interaction of the F-box receptors with IAA mediates cell and tissue-specific transcriptional responses Bargmann and Estelle (2014).

The mode of action of auxins was summarized by Machakova et al. (2008) as follows:

❖ The perception of the auxin signal

When auxins are added to culture media, they are detected by the plant cells, organs and tissues in culture. In fact, all plant cells have receptors that, after perceiving auxin signals, start a series of biochemical and molecular activities that result in a physiological reaction (Machakova et al. 2008). Such receptors have been described by many authors. For example: The Zea mays Auxin-Binding Protein ZmABP1 from maize.

❖ The signal transferring process

According to Millner (2001), plants have mechanisms responsible of the transduction of auxin signals. This is based on the regulation of the ubiquitin (i.e. a small and highly-conserved protein that enables protein degradation) conjugating pathway by auxin (Machakova et al. 2008).

❖ The physiological response of plant cells in culture

Most of the processes that are controlled by auxins implicate a signal transformation and thus a gene expression modulation (Theologis 1986). In fact, in many plant species, several auxin-responsive genes have been identified after being treated by an auxin (Machakova et al. 2008).

The physiological response of plant cells and tissues in culture depends on auxin concentration. It is well known that, increasing the concentration of auxins up to a certain level increases its effect. However, after that level, an opposite or inhibitory effect may be observed. This is due to the increase in ethylene production at a high auxin concentration (Machakova et al. 2008).

In the majority of cases, auxins show a potent effect when their concentration is lower than 10 µM. This concentration range stimulates cell division and elongation due to a change in gene expression.

Generally, auxins are used in combination with cytokinins. In fact, the combination of these two PGR families results in a more potent effect than that observed when an auxin is used alone. Besides, the combination of auxins and cytokinins regulate cell division, differentiation and growth, and affects the different developmental processes. It is important to well optimize the auxin and cytokinin concentrations to achieve the morphogenic response required.

Auxins are also known to promote differentiation of vascular bundles, maintain the polarity of the plant and its organs, and stimulate the formation of adventitious roots (Machakova et al. 2008).

ROLES OF AUXINS IN PLANT TISSUE CULTURE

ROLE OF AUXINS IN CELL DIVISION, DIFFERENTIATION AND GROWTH

When plant cells and tissues are placed in a culture medium containing PGRs, a cell division, differentiation and growth process starts. This is due to the fact that the exogenous PGRs added to the culture medium interact with the endogenous plant hormones. This process results in in vitro morphogenesis. Thus, adding PGRs to the culture medium, in particular auxins and cytokinins, is necessary for cell division and subsequent morphogenesis.

According to Lo Schiavo et al. (1989), the use of auxins in the culture medium make the DNA more methylated than in a normal state. This methylation results in the re-programming of differentiated cells. Besides, it was reported that high DNA methylation results in the induction of somatic embryogenesis (Leljak-Levanic et al. 2004).

ROLE OF AUXINS IN RHIZOGENESIS

Auxins are generally used to induce adventitious rooting. They are the most frequently used plant hormones to promote root primordium development in plants.

Cell dedifferentiation is the first step of root primordium induction. Afterwards, the dedifferentiated cells become the primordium initials. Primordium development takes place thereafter by cell division and multiplication (Haissig, 1970).

According to Haissig (1970, 1972), auxins, when used alone, do not initiate primordium formation, even though auxins stimulate primordium development. In fact, this author reported that the use of the auxin IAA does not initiate the cellular dedifferentiation that establishes the primordium initials. In fact, IAA initiates primordium formation only in the predisposed cells, but this auxin alone is not able to create the required predisposition (Haissig 1970). According to Haissig (1974), before the initiation of primordium, the cells must attain first a predisposition state to respond to endogenous or exogenous auxins. These findings may explain the differences observed in rhizogenesis when culturing different plant species, genotypes or even explants in media containing the same auxin.

It is well known that root induction in plants depends on many factors, and among these factors the most important are auxin type and concentration, and their specific impact that varies depending on plant species. The type of auxin used in in vitro culture has a significant impact on root induction, and the optimum auxin type must be chosen depending on the plant species, genotype and the explant placed in culture.

Indole-3-butyric acid and NAA are the most frequently used auxins to induce rhizogenesis in plants cultured in vitro. In *Vigna mungo*, NAA showed better rhizogenesis rates than IBA (Adlinge et al. 2014; Mony et al. 2010). However, IBA showed better results than NAA in inducing roots in shoots of *Albizia lebbek* (Perveen et al. 2011). In *Panax ginseng*, IBA was more effective in inducing rhizogenesis than NAA (Kim et al. 2003). In *Cassia angustifolia*, IBA showed better results in root induction from microcuttings than IAA and NAA (Siddique et al. 2013).

In addition to auxin type, the concentration of the auxin used in the culture medium plays an important role in inducing roots. In many plant species, low concentrations of auxins showed better results in inducing roots than high concentrations. In fact, high auxin concentrations induce callogenesis instead of rhizogenesis. For example, in *Leucaena leucocephala* (Naik et al. 2000; Shaik et al. 2009), *Caragana fruticosa* (Zhai et al. 2011) and *Cassia siamea* (Parveen et al. 2010).

While in many plant species, root induction was achieved by combining auxins and cytokinins, in many other plant species, auxins were used alone to induce rhizogenesis (Machakova et al. 2008).

ROLE OF AUXINS IN ORGANOGENESIS

It is well known that adding auxins to culture medium stimulate the initial growth of meristem and shoot tip explants. Combining an auxin at a low concentration with a cytokinin promote adventitious bud induction, and shoot bud multiplication, even though, in many plant species, the use of a cytokinin alone successfully induced organogenesis. The type and concentration of auxin to use to induce organogenesis depend strongly on the genotype and the explant type, and should be optimized in order to avoid undesirable responses such as callogenesis or rhizogenesis.

ROLE OF AUXINS IN CALLOGENESIS

Callus is a cluster of undifferentiated and disorganized cells that are induced on the surface of the explant in culture. These cells are the result of the stresses caused by culture medium components and in vitro culture conditions, mainly auxins. The induced calli grow heterotrophically with no defined structure. Generally, after induction, calli are transferred to a culture medium that will promote the formation of somatic embryos or adventitious buds. Besides, calli are use and genetic transformation and cryopreservation programs.

While low concentrations of auxins are generally used to induce rhizogenesis, high concentrations are used for callogenesis. For example, in date palm, it is well known that a high concentration of 2,4-D, of up to 100 mg/L, is used to induce callogenesis (Mazri et al. 2017). Dewir (2016) reported that, in some plant species, high auxin concentrations retard rooting and instead induce callogenesis. For example, in *Caragana fruticosa* shoots (Zhai et al. 2011). Accordingly, the concentration of auxin is a key factor to induce callogenesis.

Different types of auxins are used to induce callogenesis, depending on plant species, genotype, explant type and the associated PGRs. According to Machakova et al. (2008), auxins, when added to the culture medium, are able to change the previously determined differentiated state and the genetically programmed physiology of plant cells, tissues and organs in culture. According to Gaspar et al. (1996), the added auxins interact with endogenous plant hormones, which resulted in cell division, dedifferentiation and morphogenesis. In fact, plant cells start to divide and enter a dedifferentiated state. This is due to the methylated state of DNA caused by auxins, which results in a reprogramming of the differentiated cells (Lo Schiavo et al. 1989).

2,4-dichlorophenoxyacetic acid is the most frequently used auxin to induce callogenesis. However, it is well known that 2,4-D may induce somaclonal variation in plant cells and tissues in culture. Somaclonal variation is a genetic variability observed in the explants maintained in culture. This genetic variability might be undesirable if the purpose is the production of true-to-type plants. Thus, in many plant species, other auxins such as NAA, NOA, IAA or picloram were used to replace 2,4-D in callogenesis induction (Mazri et al. 2017). Sometimes, callogenesis is initiated in a culture medium supplemented with 2,4-D, then calli are transferred to a medium containing another auxin or to a PGR-free medium in order to produce somatic embryos or adventitious buds.

In the majority of cases, auxins are combined with cytokinins in the culture medium in order to induce callogenesis. However, in many other cases, auxins were used alone and successfully induced callus formation. For example, the auxin 2,4-D. In some plant species, a high concentration of auxins is needed to induce callogenesis.

ROLES OF AUXINS IN SOMATIC EMBRYOGENESIS

In many plant species, auxins were used to induce somatic embryogenesis. In fact, after callus induction, different morphogenic responses may be observed. For example, root formation, adventitious bud formation or somatic embryo expression. At this stage, and depending on plant species, genotype and the explant type, an auxin is generally used at a low concentration than that used to induce callogenesis, in order to promote somatic embryo formation.

The auxins added to the culture medium stimulate cell dispersion, especially in suspension culture. The addition of an auxin-cytokinin combination to the culture medium promote cell aggregation. When auxins are added at high concentrations to the culture medium, they promote callogenesis and embryogenesis, on the other hand, they prevent organogenesis. The necessity to add a cytokinin with the auxin is because auxins may hamper chlorophyll formation in calli and cells in suspension, while cytokinin promote it (Machakova et al. 2008). Thus these two PGRs have a complementary role. In fact, in many plant species, the use of 2,4-D to induce somatic embryogenesis showed a reduction in the chlorophyll content. Indeed, even a low concentration of 2,4-D may delay chlorophyll formation and reduce its accumulation. Besides, reducing 2,4-D concentration or transferring calli to an auxin-free medium showed an increase in chlorophyll content and the beginning of callus greening. Other auxins such as IAA and NAA were also reported to reduce the chlorophyll content in embryogenic cultures, but they have a less potent effect than 2,4-D.

USE OF AUXINS IN PLANT TISSUE CULTURE

Auxins are generally employed to achieve plant regeneration through the different techniques of tissue culture. The concentration and type of auxin to use depend strongly on the genotype, the explant type and the technique used. According to Machakova et al. (2008), the type and concentration of auxin that should be added to the culture medium depend on the following factors:

- ❖ The concentration of endogenous plant hormones.
- ❖ The interaction between exogenous and endogenous auxins.
- ❖ The sensitivity of plant tissues to auxins.
- ❖ The level to which the exogenous auxin is used or transported to the target plant tissue.
- ❖ The inactivation of auxins in the culture medium and within the plant material in culture.
- ❖ The in vitro regeneration technique.

In the following subsections, we will give some examples that highlight the significant impact of auxins in in vitro regeneration of some plant species, by employing the most frequently used in vitro culture techniques: somatic embryogenesis, organogenesis and microcuttings.

USE OF AUXINS FOR MICROPROPAGATION THROUGH SOMATIC EMBRYOGENESIS

Plant tissues, cells and organs are characterized by totipotency, which the ability to produce complete plants from a single cell under appropriate culture conditions. Thus, Plant tissues are able to differentiate and form embryos from somatic cells without pollination as a reaction to the different stresses that are caused by culture conditions.

Somatic embryogenesis could be defined as the in vitro culture system by which a single cell or multiple cells placed in vitro under appropriate culture conditions go through different physiological and biochemical events and processes to modify their own ontogenic program. This will result in the formation of a bipolar structure morphologically similar to zygotic embryos, and thus contains both shoot and root meristems. This bipolar structure is called 'somatic embryo'. The somatic embryo has no vascular connection with the mother explant. Under appropriate conditions (i.e. mineral salts of the culture medium, PGRs, carbon source, medium texture, light conditions …), somatic embryos are able to germinate and convert into complete plants.

There are two different patterns or pathways of the somatic embryogenesis process:

- ❖ The first pattern is known as direct somatic embryogenesis. In this case, globular somatic embryos are formed directly on the surface of the explant in culture. Accordingly, there is no intermediate development of calli (callogenesis).
- ❖ The second pattern is known as indirect somatic embryogenesis. In this case, calli are first induced on the surface of the explant in culture, which is known as callogenesis. Afterwards, calli are transferred onto a culture medium that will stimulate their proliferation and then they are placed on another culture medium for somatic embryogenesis expression. Indirect somatic embryogenesis is the most used pattern of this regeneration system. Auxins are generally added to culture medium to induce callogenesis. The type and concentration of auxin depend on plant species, genotype and explant type.

❖ It is worth noting that, in addition to direct and indirect somatic embryogenesis, there is another pattern of somatic embryogenesis known as secondary somatic embryogenesis. In this case, new somatic embryos are formed and proliferated on the previously induced somatic embryos that are placed on a culture medium for germination and conversion into complete plantlets.

To develop a successful regeneration system through somatic embryogenesis for a given plant species, it is important to take into account many factors such as the genotype, the age and the physiological state of the mother plant, the explant characteristics, including the age and the size, the disinfectants used for surface-sterilization and the time of disinfection, the concentrations of the mineral salts of the culture medium, the texture of the culture medium (liquid agitated or seimi-solid), carbon source type and concentration, light conditions and PGRs, most importantly, auxins. In fact, it is well known that auxins play a crucial role in somatic embryogenesis and that they are used in higher concentrations than cytokinins. In some plant species, the use of auxins alone was sufficient to induce callogenesis and somatic embryogenesis, and to promote somatic embryo germination and conversion into complete plants.

Following are the different culture phases of the indirect somatic embryogenesis process:

❖ Callus induction: In this first phase of culture, explants are placed on a culture medium that will promote callus formation. Generally, auxins are added to the culture medium, either alone or in combination with a cytokinin to induce callogenesis. Callus appears first in the cut end of explants in culture before invading the whole explant surface. Depending on many factors such as the species, the genotype, the explant type and the other culture conditions, the formed calli may have different morphological aspects. After induction, calli may be transferred to another culture medium for proliferation. Afterwards, calli are placed onto the expression medium, which is the beginning of the second culture phase of the somatic embryogenesis process.

❖ Somatic embryo expression: In this culture phase, the induced calli are transferred to the expression medium in which globular somatic embryos will arise. In many plants species, researchers have used a PGR-free culture medium or a medium with lower PGR concentrations than those of the induction phase. After their emergence, globular somatic embryos may be proliferated either on the same culture medium or on another one with a different composition.

❖ Somatic embryo maturation: During this phase of culture, globular somatic embryos previously formed on the expression medium are transferred to another medium that will stimulate their growth and maturation. The components and texture of the culture medium used for somatic embryo maturation depend on plant species. However, it is widely recommended to use a liquid texture with agitation, and to add ABA and an osmoticum to the maturation medium. In fact, ABA and osmotica were reported to promote the accumulation of storage proteins and lipids and the synchronization of somatic embryo maturation. The process of somatic embryo maturation varies between monocotyledons and dicotyledons. In the first case, globular somatic embryos are elongated to reach maturation while in the second case, somatic embryos go through different developmental phase to form a heart shape and a torpedo shape before reaching the dicotyledonous stage.

❖ Somatic embryo germination and conversion into complete and healthy plantlets: After maturation, somatic embryos are transferred to a culture medium that will stimulate their germination and subsequent development. In many plant species, it is recommended to add auxins to the germination medium since they will promote somatic embryo development by inducing root formation. Somatic embryo germination was also reported in PGR-free culture media, in media supplemented with an auxin-cytokinin combination or with gibberellic acid (GA_3).

❖ Plantlet acclimatization: During this phase, the somatic embryos that were successfully converted into complete and healthy plantlets are transferred to the glasshouse for acclimatization to ex vitro conditions.

Auxins have been widely used in the somatic embryogenesis process of many plant species. Mandal and Gupta (2003) evaluated the impact of different types and concentrations of auxins on somatic embryogenesis of safflower (*Carthamus tinctorius* L 'Girna'). Seeds of safflower were disinfected with 0.1% $HgCL_2$ followed by three rinses in sterile distilled water before germination in semis-solid and PGR-free MS medium. After germination, cotyledons were cut into 5-8 mm long segments and were used as explants. The explants were cultured on semi-solid MS medium supplemented with different auxin types and concentrations: NAA (0, 1.07, 2.69, 5.37, 10.74 or 21.48 µM), IAA (0, 1.14, 2.85, 5.7, 11.42 or 22.84), 2,4-D (0, 0.9, 2.26, 4.52, 9.05 or 18.1 µM) or *p*-chloro-phenoxyacetic acid (*p*-CPA; 0, 1.07, 2.68, 10.72 or 21.44 µM). These auxins were used either alone or in combination with 6-benzylaminopurine (BAP). The germinated embryos were cultured on semi-solid MS/2 medium supplemented with different concentrations of NAA (0.54, 1.07 or 2.69 µM) or IBA (0.49, 0.98 or 2.46 µM).

Mandal and Gupta (2003) reported that the embryogenic frequency and the number of somatic embryos depended on auxin type and concentration, and that NAA at 10.74 µM (2 mg/L) was the optimal auxin type and concentration for high frequency of somatic embryogenesis. On the other hand, IAA provided the maximum number of somatic embryos per responding culture. Somatic embryo development also influenced by auxin type and concentration. The findings showed that the highest number of well-developed somatic embryos at the cotyledonary stage was obtained when the combination of 5.37 µM (1 mg/L) NAA and 12.22 µM (0.5 mg /L) BAP was used. Somatic embryo conversion into complete plantlets was achieved on MS/2 medium supplemented with 1.07 µM NAA.

Przetakiewicz et al. (2003) assessed the effects of different types of auxins, used either alone or in combination, on somatic embryogenesis and plant regeneration in three cereal species. In their study, the authors isolated immature embryos from immature seeds of plants cultivated in growth chambers of five cultivars of wheat, eight cultivars of barley and three cultivars of triticale. The seeds were collected after anthesis then were disinfected with 0.1% $HgCl_2$ containing one drop of Tween for 3–5 min, followed by washes with sterile distilled water. Immature embryos were then collected and cultured on the basal formulation of MS medium modified with B5 vitamins. The medium was supplemented with either 3 mg/L 2,4-D, picloram or dicamba, or their combinations at various concentrations (from 1 to 1.5 mg/L each). After 4 to 5 weeks of culture, embryogenic calli were transferred to either PGR-free medium or to a medium supplemented with the combination of 0.2 mg/L IAA and 1 mg/L BAP. The regenerants were rooted on MS/2 medium.

The results reported by Przetakiewicz et al. (2003) highlighted the significant effect of auxin and plant species and genotype on somatic embryogenesis. The percentage of explants forming embryogenic calli varied from 25% for wheat to 100% for barley. The auxins used in culture media significantly influenced the average number of the plantlets regenerated per explant. The use of dicamba, either alone or in combination with 2,4-D, resulted in the best regeneration coefficients for most barley cultivars, except for cv. Scarlett, that showed the best results in culture media supplemented with picloram or 2,4-D. On the other hand, the highest values of regeneration coefficients for the triticale cultivars cv. Wanad and cv. Kargo were observed when picloram was used. Whereas for cv. Gabo, the best medium was that containing picloram and dicamba. In the five wheat cultivars, the combination of picloram and 2,4-D at low concentrations, or the use of dicamba alone, showed the best results on embryogenic callus formation. Plantlet development was better in media containing PGRs when compared to PGR-free medium.

Baker and Wetzstein (1994) studied the influence of different types and concentrations of auxins on peanut somatic embryogenesis. Immature pegs of peanut plants were collected from the greenhouse, surface-disinfected by soaking them in a soapy water for 1 h, followed by 95% EtOH for 2 min, 3.1% NaOCl (60% commercial bleach with 5.25% NaOCl) for 10 min, then by three rinses in sterilized water. Cotyledons were then excised from embryos and cultured on the induction medium for 30 days. The induction medium consisted of MS basal formulation modified with B5 vitamins. Different PGR combinations/concentrations were added to induction media: 5, 10, 20 or 40 mg/L 2,4-D; the combination of 5 mg/L 2,4-D and 0.2 mg/L kinetin; 40, 50 or 60 mg/L 2,4-D; 20, 30, 40 or 50 mg/L NAA. Explants were then transferred to the same induction medium but without PGRs.

The results reported by Baker and Wetzstein (1994) showed a somatic embryogenesis rate ranging from 31 to 94%. In addition, it was observed that increasing the auxin concentration decreased the percentage of somatic embryogenesis and increased explant browning. However, high concentrations of 2,4-D, equal or higher than 40 mg/L, resulted in dense masses of embryogenic areas and enhanced repetitive embryogenesis. It was also observed that increasing auxin concentrations decreased precocious embryo germination. The results of this investigation highlighted the different effects of 2,4-D and NAA on peanut somatic embryogenesis. In fact, 2,4-D showed greater somatic embryogenesis percentages and higher average number of embryos. Besides, the somatic embryos induced on NAA were less succulent, harder and less pliant.

Artunduaga et al. (1988) evaluated the effects of auxin concentrations on induction and growth of embryogenic callus from two grass species: two genotypes of Old World bluestem (*Bothriochloa* spp.) and three genotypes of bermuda (*Cynodon* spp.) grasses. Shoots with immature inflorescences were collected at the beginning of the blooming season. The shoots were disinfected by using 10% commercial clorox for 2 min, followed by 70% ethanol for 1 min, then by five rinses in sterile distilled water. Immature inflorescences ($\leq$9 mm in length) were excised aseptically then cultured on MS/2 medium supplemented with 0, 1, 3, or 5 mg/L 2,4-D and 1 mg/L IAA for 8 weeks, calli were transferred to MS/2 medium supplemented with 0.5 mg/L 2,4-D and 1 mg/L zeatin to induce somatic embryo germination and their subsequent conversion into complete plantlets. The regenerated plantlets were transferred to basal medium to grow, then acclimatized.

The results found by Artunduaga et al. (1988) showed that the explants of all genotypes produced calli by the end of the fourth week of culture. Calli became embryogenic within 8 weeks after transferring them onto fresh media under a 16 h photoperiod and 25°C temperature. The use of 3 mg/L 2,4-D maximized the production of embryogenic calli in both bluestem genotypes and in two bermudagrass genotypes.

Filippov et al. (2006) studied the impact of different factors, including auxins and time of exposure to auxins, on somatic embryogenesis induction and plant regeneration from mature embryos of Russian spring and winter genotypes of wheat. Mature grains were collected from plants growing in the greenhouse. They were surface-disinfected with 96% ethanol for 1 min, followed by 15 min in 100% commercial bleach containing a few drops of Tween at constant stirring, and rinsed four times with sterile distilled water. The seeds were then imbibed in sterile water for 3 h at room temperature. Embryo-derived explants were cultured on MS medium supplemented with different auxin types and concentrations, namely 2,4-D, dicamba and 2,4,5-trichlorophenoxyacetic acid (2,4,5-T) at 6, 8, 10, 12 and 14 mg/L. In addition, the effects of time exposure to auxin on somatic embryogenesis was evaluated. The Explants were thus cultured for 7, 14, 21 and 28 days on MS medium supplemented with 12 mg/L dicamba before transferring them to PGR-free MS medium for regeneration. In a third experiment, the explants were cultured on MS medium supplemented with 12 mg/L dicamba and various concentrations of the auxins IAA, NAA or IBA (0.1, 0.5, 1.0 and 5.0 mg/L) for 3 weeks before transferring them to PGR-free MS medium for regeneration.

The results of Filippov et al. (2006) showed that all auxins were able to induce callus from explants with high frequencies, ranging from 98 to 100%. However, dicamba was more effective in inducing embryogenic calli, with a range of 21.8-38.3%. The concentration of 12 mg/L of dicamba was the one that produced the maximum of embryogenic calli as well as the high number of regenerated plants. Besides, it was found that the time exposure to dicamba had also a significant impact on somatic embryogenesis. In fact, increasing the time of exposure to dicamba from 7 to 21 days increased the rate of somatic embryogenesis and the number of regenerated plants per embryogenic callus. Adding other auxins to the induction medium, namely IAA, IBA and NAA, resulted in an increase in the number of the rate of somatic embryogenesis induction. However, the average number of regenerated shoots did not change. The best rate of somatic embryogenesis (61%) was observed when the combination of 0.5 mg/L IAA and 12 mg/L Dicamba was used.

Vondrakova et al. (2011) used embryogenic suspensor mass (ESM) of *Abies alba* Mill., derived from immature zygotic embryos for somatic embryogenesis. They isolated seeds from strobilus and rinsed them in ethanol, then they surface-sterilized them with sodium hypochlorite solution (20%) for 30 min. The embryo explants were cultured on MS/2 medium supplemented with 2 μM BAP and 2 μM kinetin to induce ESM. Embryogenic suspensor masses were then cultured on two different media to evaluate their effects on ESM proliferation. The first one was an auxin-free MS/2 medium supplemented with 2 μM BAP and 2 μM kinetin, while the second one was supplemented by 2,4–D. For maturation, ESM were cultured on MS/2 medium supplemented with 20 μM ABA, 4% maltose and 3.75% polyethylene glycol (PEG)-4000, and the effect of auxin was tested (0.25 μM 2,4-D).

The findings of Vondrakova et al. (2011) showed that the endogenous level of IAA was significantly higher when the ESM were cultured on the medium containing 2,4-D. The subsequent development of embryos was promoted by a decrease in the endogenous level of IAA in the first week of maturation. Besides, suppression of IAA synthesis by an auxin inhibitor did not stimulate the development of embryos. Somatic embryo maturation (development from the globular to the cotyledonary stage) was correlated with an increase of the endogenous auxin in the ESM. Interestingly, Early somatic embryo proliferation on a medium containing auxin was correlated with successful maturation. Besides, it was revealed that exogenous auxin treatment during maturation did not compensate for the auxin deficiency during proliferation.

In date palm, the promotive effects of auxins on somatic embryogenesis was reported by many authors. Mazri et al. (2017) used adventitious buds, leaves (distal and proximal segments) and roots to induce somatic embryogenesis. The explants were cut into 0.5 cm length segments then cultured on semi-solid MS medium supplemented with the cytokinins 6-(γ,γ-Dimethylallylamino)purine (2iP) or kinetin at 4.5 µM and different auxins: 2,4-D, NOA, NAA or picloram at 45 µM for 4 to 6 months. For somatic embryogenesis expression, calli were transferred to the same induction medium but supplemented with only 10% of the PGR combination used.

The findings of this study showed that somatic embryogenesis of date palm is strongly affected by auxin type. In fact, despite that all auxins induced somatic embryogenesis, 2,4-D and picloram showed significantly higher somatic embryogenesis percentages than the other auxins (NOA and NAA). In fact, the use of 2,4-D and picloram resulted in somatic embryogenesis frequencies ranging from 80 to 86% in bud explants. Furthermore, 2,4-D and picloram were the only auxins to induce somatic embryogenesis in proximal leaf explants, with a frequencies ranging from 12 to 16%. Somatic embryos successfully germinated and converted to complete plants, which showed normal growth and development after acclimatization.

Mazri et al. (2018) evaluated the effects of different auxin types and concentrations on somatic embryogenesis of date palm cv. Mejhoul. Adventitious bud explants were cultured for 6 months on semi-solid MS medium supplemented with 5 µM 2iP and the auxins dicamba or picloram at different concentrations: 22.5, 45, 90, 225 or 450 µM. After induction, calli were transferred to PGR-free

MS medium for one month. Somatic embryo maturation was performed on liquid MS medium supplemented with 20 g/L PEG, then for somatic embryo germination and conversion, different concentrations of NAA and BAP (0-5 μM) were evaluated.

The results of this study showed that somatic embryogenesis induction and expression vary significantly depending on the culture medium used. The highest somatic embryogenesis percentage was 78% and was obtained on MS medium supplemented with 225 μM picloram. However, these authors reported that there was no significant difference with media containing 45, 90 and 450 μM picloram, which exhibited somatic embryogenesis induction percentages ranging from 70 to 74%. Mazri et al. (2018) reported also that adding dicamba to the culture medium resulted in very low somatic embryogenesis rates, ranging from 4 to 20%, which highlights the significant impact of auxin type in date palm somatic embryogenesis. Regarding somatic embryo germination, it varied depending on the concentrations of NAA and BAP used. In fact, the use of PGR-free MS medium showed a very low germination rate with only 8%. The highest germination rates ranged from 48 to 52%, with no significant differences, and were observed on the media containing 2.5 μM NAA and 2.5 μM BAP, 5 μM NAA and 2.5 μM BAP, 2.5 μM NAA and 5 μM BAP, and finally 5 μM NAA and 5 μM BAP. Germinated somatic embryos were successfully transplanted and a high survival rate was observed after plantlet acclimatization.

Kumar et al. (2017) examined the influence of different 2,4-D concentrations on somatic embryogenesis of *Hypoxis hemerocallidea*. The corms of *Hypoxis hemerocallidea* were washed with running tap water, cut horizontally and the outer epidermis was removed. The slices were washed again in running tap water then disinfected with absolute ethanol for 5 min, followed by 15 min in 1% Benlate then 20 min in 0.1% mercuric chloride. The corm slices were rinsed in sterile distilled water, cut into small explants then cultured for six weeks on semi-solid MS medium supplemented with different concentrations of 2,4-D: 5, 10, 15, 20 or 25 µM to induce callogenesis. For somatic embryogenesis induction, calli were cultured on semi-solid MS medium supplemented with different concentrations and combinations of 2,4-D (5, 10, 15, 20 or 25 µM) and BAP (0, 5 µM). After 6 weeks of culture, the somatic embryogenesis percentage and the mean number of somatic embryos were calculated. Embryogenic calli were then transferred to MS medium supplemented with 15 µM 2,4-D and 2.5 µM BAP for somatic embryo maturation. Somatic embryo germination was achieved on half-strength or full strength MS medium while somatic embryo conversion into plantlet was achieved on half-strength or full strength MS medium supplemented with 1.44 µM GA$_3$.

Embryogenic calli were obtained on MS medium containing 15 µM 2,4-D and 2.5 µM BAP. In fact, this combination gave the highest percentage of embryogenic calli (100%) and the highest mean number of somatic embryos (30.56). Besides, 2,4-D was essential to the proliferation and maintenance of the somatic embryo cultures. Somatic embryo conversion into complete plantlets was promoted by adding 1.44 µM GA3 to the medium and successful acclimatization was achieved under ex vitro conditions, with a high survival rate of 90%.

USE OF AUXINS FOR MICROPROPAGATION THROUGH ORGANOGENESIS

Plant cells, tissues and organs have the potential to regenerate complete plants when they are cultured in vitro under appropriate conditions (i.e. culture medium mineral salts, PGRs, carbon source, medium texture, light conditions …). As already mentioned, this ability is called totipotency. Organogenesis generally refers to adventitious organ formation under in vitro culture conditions. It is the regeneration process based on promoting the explant ability to form adventitious buds that will be cultured on different culture media in order to stimulate their multiplication, elongation and rooting.

The culture conditions, including the auxins added to the culture medium and their concentrations, play an important role in organogenesis. In fact, they create different types of stresses that result in organ formation. For example, adventitious buds, shoots and roots. These organs will have a vascular connection with the mother explant. After adventitious shoot bud induction, they will grow and develop to form a complete plant.

Adventitious buds, shoots and roots may be formed directly on the explant in culture, or after the formation of callus. Developing a successful regeneration system through organogenesis depends on many factors including the species, the genotype, the explant type, the composition of the culture medium, the culture conditions and most importantly PGRs, mainly auxins and cytokinins.

The steps of a regeneration system through organogenesis are as follows:

- ❖ Adventitious bud induction: During this culture phase, adventitious buds are induced from the explant in culture. The buds may be formed directly on the explant or after the formation of callus, depending on the species and the culture medium.
- ❖ Shoot bud multiplication: After the induction phase, the shoot buds are transferred to a multiplication medium which generally contains an auxin-cytokinin combination. The purpose of this culture phase is the proliferation of shoot buds in order to produce later the maximum number of plants.
- ❖ Shoot elongation: Using a scalpel, adventitious shoot buds are singled out and transferred to a culture medium that will promote

their elongation. Generally, the culture medium contains an auxin-cytokinin combination that will promote cell division, growth and elongation.

❖ Shoot rooting: After elongation, shoots are transferred to a culture medium that will stimulate rhizogenesis. The culture medium is generally supplemented with an auxin or the combination of more than one auxin, with or without cytokinin. In fact, in many plant species, the use of auxins alone has been proven to be sufficient for root induction.

❖ Plantlet acclimatization: During this final step, the plantlet obtained in vitro through organogenesis are acclimatized to the conditions of the glasshouse.

Auxins have been widely used for plant organogenesis. Tabei et al. (1991) used cotyledons of mature seeds, cotyledons and hypocotyls of seedlings, and leaves and petioles of young plantlets of *Cucumis melo* L. as explants to evaluate the effects of auxins on organogenesis. The seed coats were removed then the seeds were surface sterilized with 1% sodium hypochlorite for 15 min followed by three rinses in sterilized distilled water. The cotyledons were divided into two segments then cultured on semi-solid MS medium. Afterwards, the expanded cotyledons and hypocotyls of seedlings were cut into small segments before being placed in the culture medium. Regarding leaves and petioles, they were taken from young leaves of 3 week old plantlets. The explants were cultured on semi-solid MS medium supplemented with 0.1 mg/L BAP and different types and concentrations of auxins: 0.01 to 2 mg/L for 2,4-D, 0.01 to 25 mg/L for NAA and 0.01 to 300 mg/L for IAA.

The results of Tabei et al. (1991) showed that the use of 2,4-D at different concentration resulted in callus induction. Callus proliferation was more important on MS medium supplemented with 0.1 mg/L 2,4-D when compared to the other auxin types and concentrations. Adventitious shoots were formed directly on the explant, while calli resulted in the formation of somatic embryos. Mature seed cotyledons showed the highest adventitious shoot induction when the culture medium was supplemented with 0.01 mg/L 2,4-D, with an organogenesis rate of 80%. When IAA was used, the formation of adventitious shoots occurred at concentrations ranging from 0 to 3 mg/L, adventitious shoots and somatic embryos were formed at the concentration of 10 mg/L while the concentrations ranging from 25 to 100 mg/L resulted in the formation of somatic embryos. Besides, the use of IAA resulted in the formation of the lowest rate of abnormal somatic embryos. The highest rate of adventitious shoot formation was 55% on MS medium containing 1 mg/L IAA. When NAA was added to the culture medium, adventitious shoots were formed at concentrations of 0, 0.01 and 0.1 mg/L, while higher concentrations (3-10 mg/L) resulted in the formation of somatic embryos. The use of NAA at the concentration of 1 mg/L resulted in callus induction, but neither adventitious shoots nor somatic embryos were formed. Besides, the use of NAA resulted in the formation of abnormal somatic embryos in the majority of cases.

When low concentrations of auxins were used, adventitious shoots formed multiple shoot clusters. Shoot elongation was also influenced by auxin type. The majority of shoots induced on MS medium supplemented with IAA showed normal elongation and plantlet conversion. On the other hand, the use of 2,4-D showed few normally elongated shoots while those obtained on MS medium containing NAA failed to grow normally. Adventitious shoots were transferred to semi-solid MS medium supplemented with 0.2 mg/L gibberellic acid 4 (GA_4) for further development.

Bonfill et al. (2002) studied the influence of auxins on organogenesis in *Panax ginseng*. Two-year-old rhizomes of *Panax ginseng* were disinfected by immersion for 30 s in 70% ethanol and sonication for 20 min in 20% sodium hypochlorite containing Tween 20, followed by several rinses with sterile distilled water. Rhizome discs were cultured on semi-solid MS medium supplemented with 0.1 mg/L kinetin and 1 mg/L 2,4- D for callus induction. After six months of culture, calli were cultured on MS medium supplemented with 0.1 mg/L kinetin and 2 mg/L of either 2,4-D, IBA or NAA.

The results of Bonfill et al. (2002) showed that the formed calli were either non-organogenic or produced buds, roots or both. The use of 2,4-D inhibited organogenesis in calli. These non-organogenic calli were friable, spongy and light. The calli cultured on the medium containing IBA and NAA were compact and dark, and show high organogenic potential. The use of IBA resulted in the formation of a high number of buds and thin and necrotized roots. The use of NAA resulted in lower organogenic capacity, with the induction of buds and thick and well-growing roots.

Landi and Mezzetti (2006) examined the impact of different auxins, along with thidiazuron (TDZ) and genotypes, on organogenesis of *Fragaria*. Nine octoploid *Fragaria x ananassa* cultivars and breeding selections; two octoploid breeding selections from *F. virginiana glauca* inter-species crosses; two diploid *F. vesca* cultivars; and one diploid clone of *F. nubicola* Lindl were used. Lateral buds of different genotypes were surface-sterilized for 20 min in a 2% (v/v) chloride-active solution, followed by several rinses with sterile distilled water. The explants were cultured on semi-solid MS medium to induce shoots. The shoots that developed were then transferred to semi-solid MS medium supplemented with 1.1 µM BAP to promote proliferation. Afterwards, young expanded leaves were detached from 4-week-old in vitro proliferating shoots and the leaf lamina were cut transversally and cultured on semi-solid MS medium supplemented with four different PGR combinations: the cytokinin TDZ was used at the concentration of 4.54 µM either alone or in combination with three different auxins: IBA at 0.98 µM, 3-Benzo[b] selenienyl acetic acid (BSAA) at 0.84 µM or 2,4-D at 0.90 µM. The explants were cultured under either a 16h photoperiod or dark conditions.

The findings of Landi and Mezzetti (2006) showed that PGRs significantly affect the induction of calli, shoots and roots. Combining TDZ with an auxin stimulated callus proliferation. The highest shoot induction rate was observed when the medium was supplemented with the combination of TDZ and IBA. Whereas the use of BSAA resulted in the highest number of regenerated shoots from leaves and in callus production in most of the genotypes. The highest root induction rate was observed when the medium was supplemented with the combination of TDZ and BSAA, while root formation was not observed when TDZ was used alone. IBA and TDZ stimulated shoot regeneration from leaves in almost all of the genotypes evaluated, while the combination of BSAA and TDZ and that of 2,4-D and TDZ stimulated regeneration in only some genotypes.

Makunga et al. (2005) assessed the effects of auxins to develop an in vitro regeneration system for *Thapsia garganica* through direct organogenesis. The leaves of stock plants of *T. garganica* growing in a sand:soil mixture in greenhouse for five growth seasons were collected, washed and disinfected for 5 min in 70% ethanol, followed by incubation in 1% Benlate for 10 min before surface-decontamination for 15 min in 1% NaOCl that was followed by three rinses with sterile water. Petiole and leaflet explants were cultured on semi-solid MS medium supplemented with different combinations of auxins and cytokinins to evaluate their effects on differentiation: The auxins NAA, IAA and 2,4-D and the cytokinins BAP and kinetin at concentrations ranging from 0 to 2 mg/L.

Afterwards, the explants with visible response were transferred to MS medium supplemented with 1 mg/L IAA and 1 mg/L BAP or 1 mg/L IAA and 2 mg/L BAP or 1 mg/L IAA and 2 mg/L kinetin or 2 mg/L 2,4-D and 1 mg/L BAP or 2 mg/L kinetin. The developed shoots were transferred to a growth medium supplemented with 1 mg/L IAA and 1 mg/L BAP or 1 mg/L IAA and 2 mg/L BAP or 1 mg/L IAA and 2 mg/L kinetin or 2 mg/L kinetin. Healthy shoots (5 cm or more) were singled out and transferred to semi-solid MS medium supplemented with combination of 0.5 mg/L NAA and 1.5 mg/L BAP or 1 mg/L NAA and 3 mg/L BAP or 2 mg/L kinetin. Afterwards, semi-solid MS media supplemented with three different auxins, namely IAA, NAA, and IBA, at the concentration of 1 mg/L, along with a non-auxin treatment, were evaluated for rooting.

The results of Makunga et al. (2005) showed that the combination of 0.5 mg/L NAA and 1.5 mg/L BAP was comparable for direct shoot organogenesis to that of 2 mg/L kinetin. Besides, these two media were the only able to induce direct in vitro regeneration from all explant types without callus formation. Subculturing the induced shoots onto MS medium supplemented with the combination of 0.5 mg/L NAA and 1.5 mg/L BAP for a second cycle has promoted indirect organogenesis. The Other combinations showed the formation of calli that preceded adventitious shoot induction or took place at the same time as adventitious shoot induction.

USE OF AUXINS IN PROPAGATION THROUGH MICROCUTTINGS

Propagation through microcutting (i.e. micropropagation by nodal stem cutting or through segmentation of elongated shoots) is a vegetative plant micropropagation method based on the multiplication of microcuttings containing axillary buds. The microcuttings are obtained by dividing semi-hardwood cuttings into small segments and then culturing them in vitro. The steps of this in vitro multiplication technique are as follows:

- ❖ Semi-hardwood cuttings of 10 to 20 cm in length are taken from healthy mother plants that are growing in the greenhouse or in open field.
- ❖ The semi-hardwood cuttings are introduced to the laboratory and thoroughly washed with tap water.
- ❖ The semi-hardwood cuttings are surface-disinfected under a laminar flow hood then washed for three to five times with sterile distilled water.
- ❖ The semi-hardwood cuttings are cut into small segments of 1.5 to 3 cm in length and that comprise a single node or double nodes.
- ❖ The microcuttings are cultured on a culture medium and under suitable in vitro culture conditions in order to promote the growth and development of axillary shoots.
- ❖ The developed shoots will be divided into small segments that will be transferred to a culture medium in order to promote their growth and elongation.
- ❖ This step can be repeated as much as needed in order to maximize plant production.
- ❖ Subcultures are made every 30 to 60 days onto the same fresh culture medium.
- ❖ The shoots are transferred to rooting medium, which generally contains auxins.
- ❖ Similarly to the somatic embryogenesis and organogenesis processes, the final step of micropropagation through microcuttings consists of transferring the rooted plants to the glasshouse for acclimatization to ex vitro conditions.

Baraldi et al. (1995) evaluated the relationship between the concentrations of the endogenous auxin IAA and polyamines (PAs) and the metabolism of IBA with the process of in vitro rooting in two *Pyrus communis* cultivars: cv. conference, which is known to be easy to root, and doyenne d'hiver, which is known to be difficult to root. Single shoots of 15 mm length from these two cultivars were taken from proliferating cultures and transferred to semi-solid MS/2 medium modified with Linsmaier and Skoog (1965) vitamins and supplemented with various concentrations of IBA and IAA: 0, 0.5, 1.5, 5 or 15 µM. The shoots were cultured under dark conditions for 7 days then transferred to a 16h photoperiod. The effect of culturing time in IBA on rooting was evaluated. Accordingly, single shoots were cultured on a medium supplemented with 1.5 µM IBA for 1, 2, 3, 4, 5 and 21 days. Afterwards, the shoots were transferred to an IBA-free medium for 21 days. As control, shoots were also cultured on an IBA-free medium. These authors also measured the endogenous auxins and PAs in the shoots that were cultured in the dark in the rooting medium supplemented with 1.5 µM IBA for 7 days. Besides, the effects of the high IBA concentrations on auxin metabolism were evaluated.

The results of this investigation showed that pear cv. doyenne d'hiver needs about a 10 times higher concentration of IBA to reach the same rooting frequency as cv. conference. Besides, an exposure time in IBA of one to two days was sufficient to promote in vitro rooting. However, the rooting efficiency varied between the two cultivars evaluated. Increasing the time of exposure of shoots in auxin-containing medium significantly enhanced the mean number of roots in cv. conference. However, this inhibited the elongation of roots in both the cultivars. The findings of this research showed also that the metabolism of IBA in both the pear cultivars did not significantly vary when IBA was used at a high level to promote rooting in cv. doyenne d'hiver.

Baraldi et al. (1995) reported that IBA was mainly conjugated into IBA glucose, while a small amount of this auxin was converted into free IAA in both the cultivars evaluated, even though in cv. doyenne d'hiver, this metabolic pathway was active only when IBA was used at a high IBA concentration. The different effects of IBA on pear cultivars were also observed on callogenesis. In fact, at a high IBA concentration, more callus was induced in cv. doyenne d'hiver than in cv. conference, which showed that the cells of these two cultivars react differently to IBA. A possible relationship between the reported processes and an early increase followed by a decrease of free IAA was observed in cv. conference. Besides, a significant increase in IAA conjugates and free putrescine was observed in cv. doyenne d'hiver. The authors suggest that the higher putrescine level may be related to the lower amount of root development, and that the differences in the uptake and metabolism of the applied auxins influence the rooting potential and the subsequent development of adventitious roots in microcuttings of *Pyrus communis*.

CONCLUSIONS

Auxins is a group of plant growth regulators widely used for plant micropropagation. Since their discovery, auxins have played a key role in plant propagation and regeneration through different techniques such as organogenesis, direct and indirect somatic embryogenesis and microcuttings. Besides, auxins are the main PGR family used for rhizogenesis.

Auxins are involved in cell division, differentiation and growth, and are fundamental for in vitro morphogenesis. They act by interacting with the endogenous plant hormones and thus modifying their levels. Consequently, morphogenic responses are observed.

Auxins are also known for their role in maintaining apical dominance and plant polarity, and are involved in the plant signaling systems, which allows plant cells to interact among them in response to different external factors and culture conditions.

With these numerous roles and functions, auxins, along with cytokinins, are now the major families of plant growth regulators. Auxins are applied in almost all the culture phases of the different regeneration systems and have been involved in the majority of achievements made in this biotechnological field of research.

REFERENCES

- Adlinge PM, Samal KC, Kumara Swamy RV, Rout GR (2014) Rapid in vitro plant regeneration of black Gram (*Vigna mungo* L. Hepper) Var. Sarala, an important legume crop. Proc Natl Acad Sci. India, Sect B Biol Sci 84:823-827

- Amissah JN, Bassuk NL (2004) Clonal propagation of *Quercus* spp. using a container layering technique. J Environ Hort 22:80–84

- Artunduaga IR, Taliaferro CM, Johnson BL (1988) Effects of auxin concentration on induction and growth of embryogenic callus from young inflorescence explants of Old World bluestem (*Bothriochloa* spp.) and bermuda (*Cynodon* spp.) grasses. Plant Cell Tissue Org Cult. 12: 13–19

- Baker CM, Wetzstein HY (1994) Influence of auxin type and concentration on peanut somatic embryogenesis. Plant Cell Tissue Org Cult. 36: 361–368

- Baraldi R, Bertazza G, Bregoli AM, Fasolo F, Rotondi A, Predieri S, Serafini-Fracassini D, Slovin JP, Cohen JD (1995) Auxins and polyamines in relation to differential in vitro root induction on microcuttings of two pear cultivars. J Plant Growth Regul 14: 49–59

- Bargmann BOR, Estelle M (2014) Auxin perception: in the IAA of the beholder. Physiol Plant 151: 52–61

- Blythe EK, Sibley JL, Tilt KM, Ruter JM (2007) Methods of auxin application in cutting propagation: A review of 70 years of scientific discovery and commercial practice. J Environ Hortic 25:166–185

- Bonfill M, Cusido RM, Palazon J, Pinol MT, Morales C (2002) Influence of auxins on organogenesis and ginsenoside production in *Panax ginseng* calluses. Plant Cell Tissue and Organ Culture 68:73–78

- Chadwick LC, Kiplinger DC (1938) The effect of synthetic growth substances on the rooting and subsequent growth of ornamental plants. Proc Amer Soc Hort Sci 36:809–816

- Chong C, Hamersma B (1995a) Automobile radiator antifreeze and windshield washer fluid as IBA carriers for rooting woody cuttings. HortSci 30:363–365

- Chong C, Hamersma B (1995b) Inexpensive IBA root-promoting solutions. Comb Proc Intl Plant Prop Soc 45:483–487

- Dewir YH, Murthy HN, Ammar MH, Alghamdi SS, Al-Suhaibani NA, Alsadon AA, Paek KY (2016) In vitro rooting of leguminous plants: difficulties, alternatives, and strategies for improvement. Hortic Environ Biotechnol 57:311–322

- Dirr MA (1981) Rooting compounds and their use in plant propagation. Comb Proc Intl Plant Prop Soc 31:472–479

- Dirr MA, Heuser CW Jr (1987) The reference manual of woody plant propagation: From seed to tissue culture. Varsity Press, Inc., Athens, GA

- Engvild KC (1985) Pollen irradiation and possible gene transfer in Nicotiana species. Theor Appl Genet 69: 457–461

- Fillippov M, Miroshnichenko D, Vernikovskaya D, Dolgov S (2006) The effect of auxins, time exposure to auxin and genotypes on somatic embryogenesis from mature embryos of wheat. Plant Cell Tissue Org Cult 84:213–222

- Friml J (2003) Auxin transport–shaping the plant. Curr Opin Plant Biol 6: 7–12

- Gaspar T, Kevers C, Penel C, Greppin H, Reid DM, Thorpe TA (1996) Plant hormones and plant growth regulators in plant tissue culture. In Vitro Cell Dev Biol Plant 32:272–289

- Haissig BE (1970) Influence of indole-3-acetic acid on adventitious root primordia of brittle willow. Planta 95:27–35

- Haissig BE (1972) Meristematic activity during adventitious root primordium development. Influences of endogenous auxin and applied gibberellic acid. Plant Physiol. 49:886–892

- Haissig BE (1974) Influences of auxins and auxin synergists on adventitious root primordium initiation and development. N Z J For Sci 4:311–323

- Hartmann HT, Kester DE, Davies FT Jr, Geneve RL (2002) Hartmann and Kester's plant propagation: Principles and Practices. 7th ed. Prentice Hall, Upper Saddle River, NJ

- Howard BH, Harrison-Murray RS, Mackenzie KAD (1984) Rooting responses to wounding winter cuttings of M.26 apple rootstock. J Hort Sci 59:131–139

- Kim YS, Hahn EJ, Yeung EC, Paek KY (2003) Lateral root development and saponin accumulation as affected by IBA or NAA in adventitious root cultures of *Panax ginseng* CA Meyer. In Vitro Cell Develop Biol 39:245–249

- Kogl F, Haagen-Smit AJ, Erxleben H (1934) Über ein neues Auxin ("Heteroauxin") aus Harn. Zeits. Physiol Chem 228:90-103

- Kumar V, Moyo M, Van Staden J (2017) Somatic embryogenesis in *Hypoxis hemerocallidea*: an important African medicinal plant. South Afr J Bot 108:331–336

- Landi L, Mezzetti B (2006) TDZ, auxin and genotype effects on leaf organogenesis in *Fragaria*. Plant Cell Rep 25:281–288

- Leljak-Levanic D, Bauer N, Mihaljevic S, Jelaska S (2004) Changes in DNA methylation during somatic embryogenesis in *Cucurbita pepo* L. Plant Cell Rep 23:120-127

- Libbenga KR, Mennes AM (1995) Hormone binding and signal transduction. pp. 272-297 in Davies P.J. (ed.) Plant Hormones: Physiology, Biochemistry and Molecular Biology. 2nd edn., Kluwer Academic Publishers, Dordrecht, Boston, London.

- Linsmaler E, Skoog F (1965) Organic growth factor requirements of tobacco tissue cultures. Physiol Plant 18:100-127

- Ljung K (2014) Auxin – a simple compound with a profound effect on plant development. Physiol Plant 151: 1–2

- Lo Schiavo F, Pitto L, Giuliano G, Torti G, Nutironchi V, Marazziti D, Vergara R, Orselli S, Terzi M (1989) DNA methylation of embryogenic carrot cell cultures and its variations as caused by mutation, differentiation, hormones and hypomethylating drugs. Theor Appl Genet 77:325-331

- Lowenfels A (1966) Various types and strengths of hormones from U.S.A., England and Holland. Comb. Proc. Intl Plant Prop Soc 16:260–263

- Ludwig-Müller J, Epstein E (1994) Indole-3-butyric acid in *Arabidopsis thaliana* III. In vivo biosynthesis. Plant Growth Regul 14:7–14

- Ludwig-Müller J, Sass S, Sutter EG, Wodner M, Epstein E (1993) Indole-3-butyric acid in *Arabidopsis thaliana* I. Identification and quantification. Plant Growth Regul 13:179–187

- Macdonald B (1987) Practical woody plant propagation for nursery growers. Timber Press, Portland, OR

- Machakova I, Zazimalova E, George EF (2008) Plant growth regulators I: introductions, auxins, their analogous and inhibitors. In: George EF, Hall MA, De Klerk GJ (eds) Plant propagation by tissue culture, vol 1, 3rd edn. Springer, Dordrecht, pp 175–204

- Makunga NP, Jager AK, van Staden J (2005) An improved system for the in vitro regeneration of *Thapsia garganica* via direct organogenesis—

influence of auxins and cytokinins. Plant Cell Tissue Organ Cult 82:271–280

- Mandal AKA, Gupta SD (2003) Somatic embryogenesis of sunflower: influence of auxin and ontogeny of somatic embryos. Plant Cell Tissue Organ Cul 72:27–31

- Mazri MA, Belkoura I, Meziani R, Mokhless B, Nour S (2017) Somatic embryogenesis from bud and leaf explants of date palm (*Phoenix dactylifera* L.) cv. Najda. 3 Biotech 7:58

- Mazri MA, Meziani R, Belkoura I, Mokhless B, Nour S (2018) A combined pathway of organogenesis and somatic embryogenesis for an efficient large-scale propagation in date palm (*Phoenix dactylifera* L.) cv. Mejhoul. 3 Biotech 8:215

- Millner PA (2001) Heterotrimeric G-proteins in plant cell signalling. New Phytol 151:165-174.

- Mony SA, Haque MdS, Alam MdM, Hasanuzzaman M, Nahar K (2010) Regeneration of black gram (*Vigna mungo* L.) on changes of hormonal condition. Not Bot Hortic Agrobot Cluj 38:140-145

- Naik SK, Pattnaik S, Chand PK (2000) In vitro propagation of pomegranate (*Punica granatum* L. cv. Ganesh) through axillary shoot proliferation from nodal segments of mature tree. Sci Hortic 79:175-183

- Parveen S, Shahzad A, Saema S (2010) In vitro plant regeneration system for *Cassia siamea* Lam., a leguminous tree of economic importance. Agrofor Syst 80:109-116

- Perveen S, Varshney A, Anis M, Aref IM (2011) Influence of cytokinins, basal media and pH on adventitious shoot regeneration from excised root cultures of *Albizia lebbeck*. J Forest Res 22:47-52

- Przetakiewicz A, Orczyk W, Nadolska-Orczyk A (2003) The effect of auxin on plant regeneration of wheat, barley and triticale. Plant Cell Tissue Org Cult 73:245–256

- Robbins JA, Campidonica MJ, Burger DW (1988) Chemical and biological stability of indole-3-butyric acid (IBA) after long-term storage at selected temperatures and light regimes. J Environ Hort 6:33–38

- Shaik NM, Arha M, Nookaraju A, Gupta SK, Shrivastava S, Yadav AK, Kulkarni PS, Abhilash OU, Rishi K (2009) Improved method of in vitro regeneration in Leucaena leucocephala. Physiol Mol Biol Plants 15:311-318

- Siddique I, Javed SB, Al-Othman M R, Anis M (2013) Stimulation of in vitro organogenesis from epicotyl explants and successive micropropagation round in *Cassia angustifolia* Vahl.: an important source of sennosides. Agrofor Syst 87:583-590

- Skoog F, Miller C (1957) Chemical regulation of growth and organ formation in plant tissues cultured in vitro. pp. 118-140 in Symp. Soc. Exptl. Biol., Number XI, The Biological Action of Growth Substances.

- Stickens D, Tao W, Verbelen JP (1996) A single cell model system to study hormone signal transduction. Plant Growth Regul 18 :149-154

- Tabei Y, Kanno T, Nishio T (1991) Regulation of organogenesis and somatic embryogenesis by auxin in melon, *Cucumis melo*. L. Plant Cell Rep. 10:225–229

- Theologis A (1986) Rapid gene regulation by auxin. Annu Rev Plant Physiol 37:407-438

- Thimann KV, Koepfli JB (1935) Identity of the growth promoting and root-forming substances of plants. Nature 135:101–102

- Van Overbeek JJ (1959) Auxins. Bot Rev 25: 269–350

- Vondrakova Z, Eliasova K, Fischerova L, Vagner M (2011) The role of auxins in somatic embryogenesis of *Abies alba*. Cent Eur J Biol 6:587–596

- Went (1928) Wuchsstoff und Wachstum. Rec Tray Bot Néerl. 25: 1
- Zhai X, Yang L, Shen H (2011) Shoot multiplication and plant regeneration in *Caragana fruticosa* (Pall.) Besser. J Forest Res 22:561-567
- Zimmerman PW, Hitchcock E (1942) Substituted phenoxy and benzoic acid growth substances and the relation of structure to physiological activity. Contr Boyce Thomps Inst 12: 321
- Zimmerman PW, Wilcoxon F (1935) Several chemical growth substances which cause initiation of roots and other responses in plants. Contrib Boyce Thomp Inst 7:209–229

Printed by Books on Demand GmbH, Norderstedt / Germany